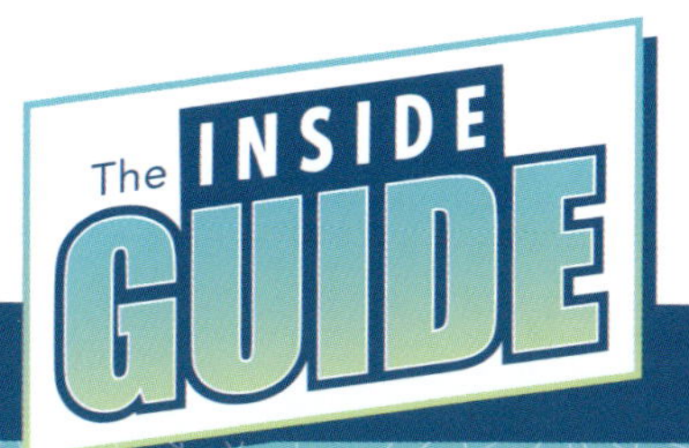

BIOLOGY BASICS

The Human Body

By Leigh McClure

Published in 2025 by Cavendish Square Publishing, LLC
2544 Clinton Street Buffalo, NY 14224

First Edition

Website: cavendishsq.com

Cataloging-in-Publication Data

Names: McClure, Leigh.
Title: The human body / Leigh McClure. Description: Buffalo, NY : Cavendish Square Publishing, 2025. |
Series: The inside guide: biology basics | Includes glossary and index.
Identifiers: ISBN 9781502673299 (pbk.) | ISBN 9781502673305 (library bound) | ISBN9781502673312 (ebook)
Subjects: LCSH: Human body–Juvenile literature. | Human biology–Juvenile literature.
Classification: LCC QP37.M28 2025 | DDC 612–dc23

Editor: Caitie McAneney
Copyeditor: Nicole Horning
Designer: Deanna Lepovich

The photographs in this book are used by permission and through the courtesy of: Cover Anatomy Image/Shutterstock.com; p. 4 MalikNalik/Shutterstock.com; p. 6 Magic mine/Shutterstock.com; p. 7 GraphicsRF.com/Shutterstock.com; p. 8 ORION PRODUCTION/Shutterstock.com; p. 9 BalanceFormCreative/Shutterstock.com; p. 10 Ekaterina Glazkova/Shutterstock.com; p. 12 Quality Stock Arts/Shutterstock.com; p. 13 Orathai Mayoeh/Shutterstock.com; p. 14 benote/Shutterstock.com; p. 15 sippakorn/Shutterstock.com; p. 16 Alter-ego/Shutterstock.com; p. 18 Ground Picture/Shutterstock.com; p. 19 Prostock-studio/Shutterstock.com; p. 20 Cat Box/Shutterstock.com; pp. 21, 27 (bicep vector) mentalmind/Shutterstock.com; p. 22 Craevschii Family/Shutterstock.com; p. 24 namaki/Shutterstock.com; p. 25 Ereny Emil Fayez/Shutterstock.com; p. 27 (brain, heart, lungs, stomach, and skull vectors) Takoyaki Tech/Shutterstock.com; p. 28 (top) vk_st/Shutterstock.com; p. 28 (bottom) fizkes/Shutterstock.com; p. 29 (top) MDV Edwards/Shutterstock.com; p. 29 (bottom) Pormezz/Shutterstock.com.

CPSIA compliance information: Batch #CWCSQ25: For further information contact Cavendish Square Publishing LLC at 1-877-980-4450.

Printed in the United States of America

CONTENTS

Understanding and caring for your body is an important part of growing up.

BODY SMARTS

The human body is an amazing machine. It's made up of **complex** systems that help you function. Your nervous system allows you to think and move. Your respiratory system keeps oxygen flowing into your body. Your digestive system ensures you get the nutrients you need to live and grow.

Our bodies don't come with instruction manuals. However, understanding the major body systems and what they need to function can help you care for the body you have. We'll start with the system that makes you who you are—the nervous system.

The Nervous System

Think of everything you've done so far today. Maybe you played soccer, ate lunch, talked to friends, or did your schoolwork. All those things are possible because of the nervous system. The nervous system controls every function of your body. It directs **voluntary** actions like running and involuntary actions like breathing.

Fast Fact

The central nervous system is made up of the brain and spinal cord. The peripheral nervous system is made up of nerves throughout the body.

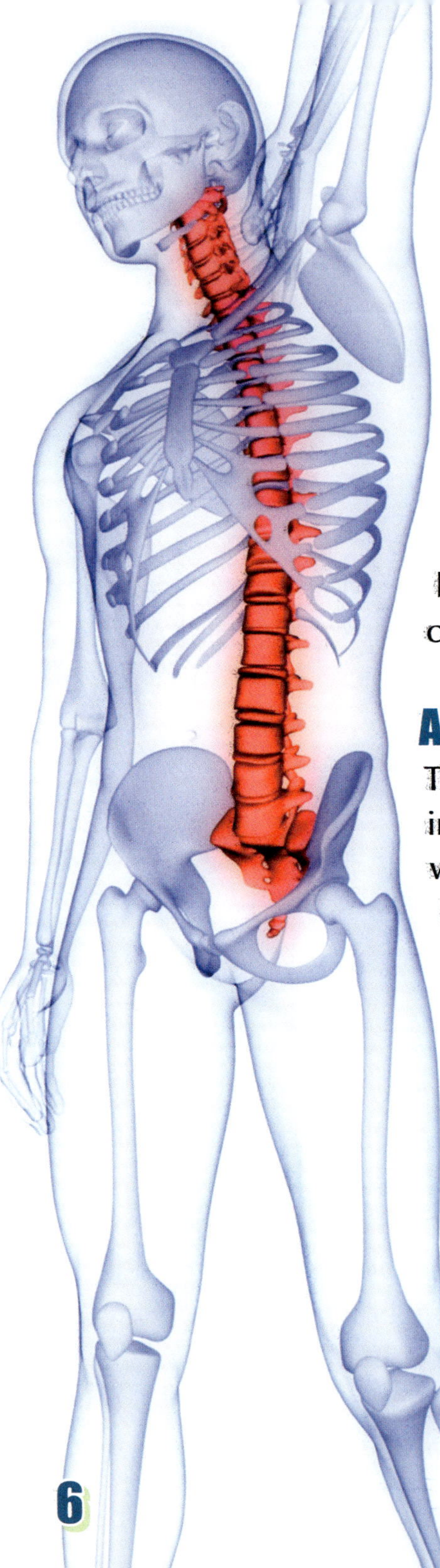

The spinal cord runs from your brain down your back. It is protected by bones called vertebrae.

The nervous system includes the brain, the spinal cord, and a network of nerves that send messages throughout the body. **Specialized** cells called neurons send messages from the brain to the spinal cord and down to the nerves in the body. They make you breathe and make your heart beat. They also allow you to solve problems, create art, and kick that soccer ball!

All About the Brain

The brain is one of the most important organs in the body. It's like a superfast computer with a huge amount of storage. In fact, the human brain has about 86 billion neurons. These neurons form connections to others over time, and the storage space just keeps growing!

The brain has different parts that control different things. The biggest part is the cerebrum. This is the part you use to reason, remember, and move your body. It controls your voluntary muscles, which help you walk, dance, and draw. There are four lobes, or parts, of the cerebrum that have their own special jobs. A smaller

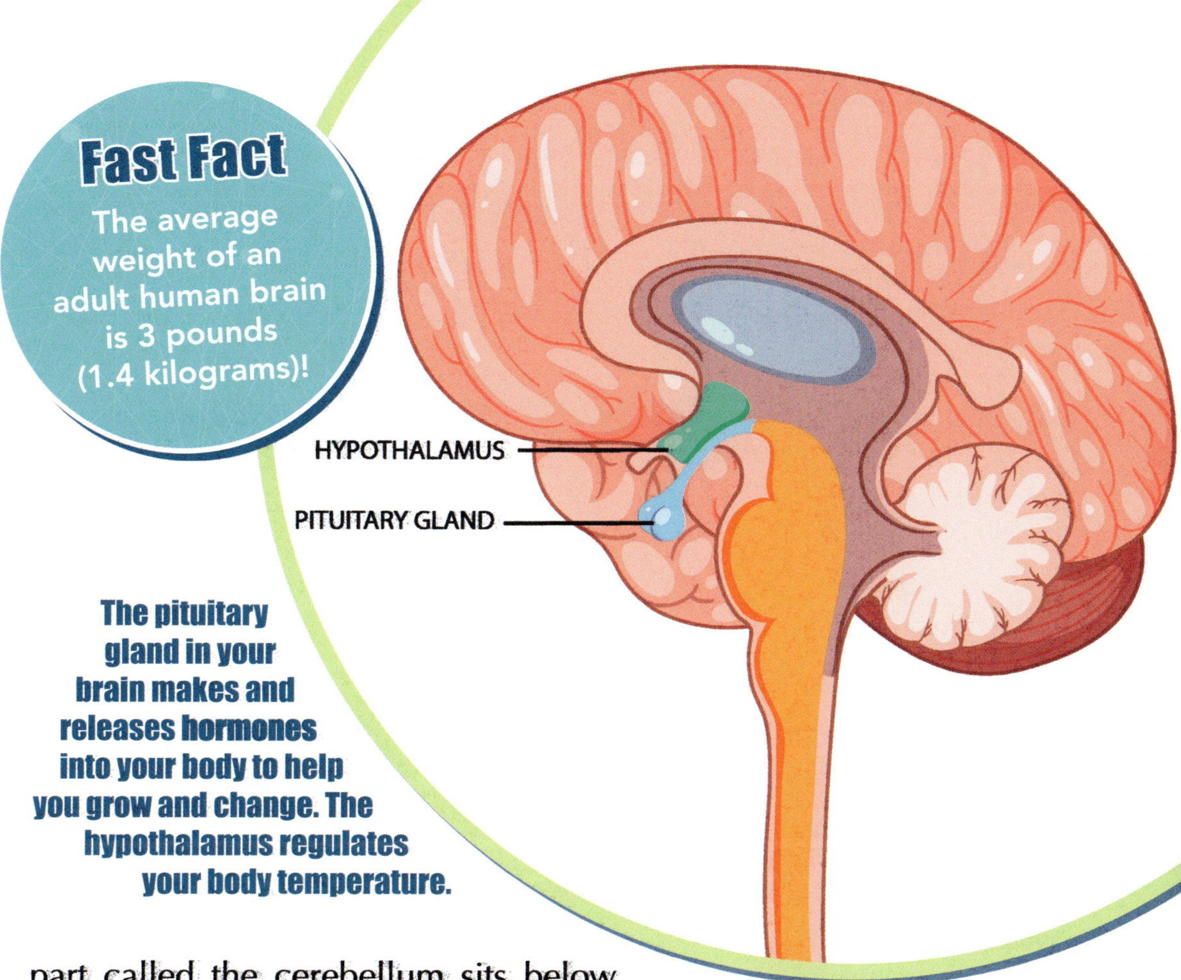

Fast Fact

The average weight of an adult human brain is 3 pounds (1.4 kilograms)!

The pituitary gland in your brain makes and releases hormones into your body to help you grow and change. The hypothalamus regulates your body temperature.

part called the cerebellum sits below the cerebrum and controls movement, **coordination**, and balance. The brain stem controls your involuntary muscles, like those in your heart and digestive system.

Sending Messages

How does the brain get the body to do what it asks? It uses nerves. The spinal cord is found down the length of your back. It's a bunch of nerves protected by a series of bones called vertebrae. The brain sends messages down the spinal column like cars on a highway. The neurons

Your somatic nervous system controls things you do on purpose, such as preparing and eating food. Your autonomic nervous system helps you digest that food.

Fast Fact

Your autonomic nervous system is like your body's autopilot. It helps you digest food, breathe, and regulate your heartbeat.

contain little branches that allow them to connect to other neurons.

Nerves connect to the brain or spinal cord and stretch throughout the body. You have sensory nerves that take in information about what you see, hear, feel, touch, and taste. They send that information to the brain. You have motor nerves that send a direction from the brain to the muscles.

NERVOUS SYSTEM ISSUES

Issues of the nervous system are called neurological issues. An injury to the brain or spinal cord can disrupt the nervous system's normal functioning. Injury to the brain can cause problems with memory, reasoning, movement, and personality. Injury to the spinal cord can cause the inability to move, or paralysis. Some diseases lead to the destruction of neurons, such as Parkinson's and Alzheimer's. This leads to trouble with movement, memory, reasoning, and more. Some diseases attack the nerves, such as multiple sclerosis (MS). In some cases, MS can make a person lose their ability to speak or walk, among other important daily functions.

Eating nutrient-rich foods like fruits and vegetables, getting enough sleep, moving the body, and doing puzzles can keep the brain healthy!

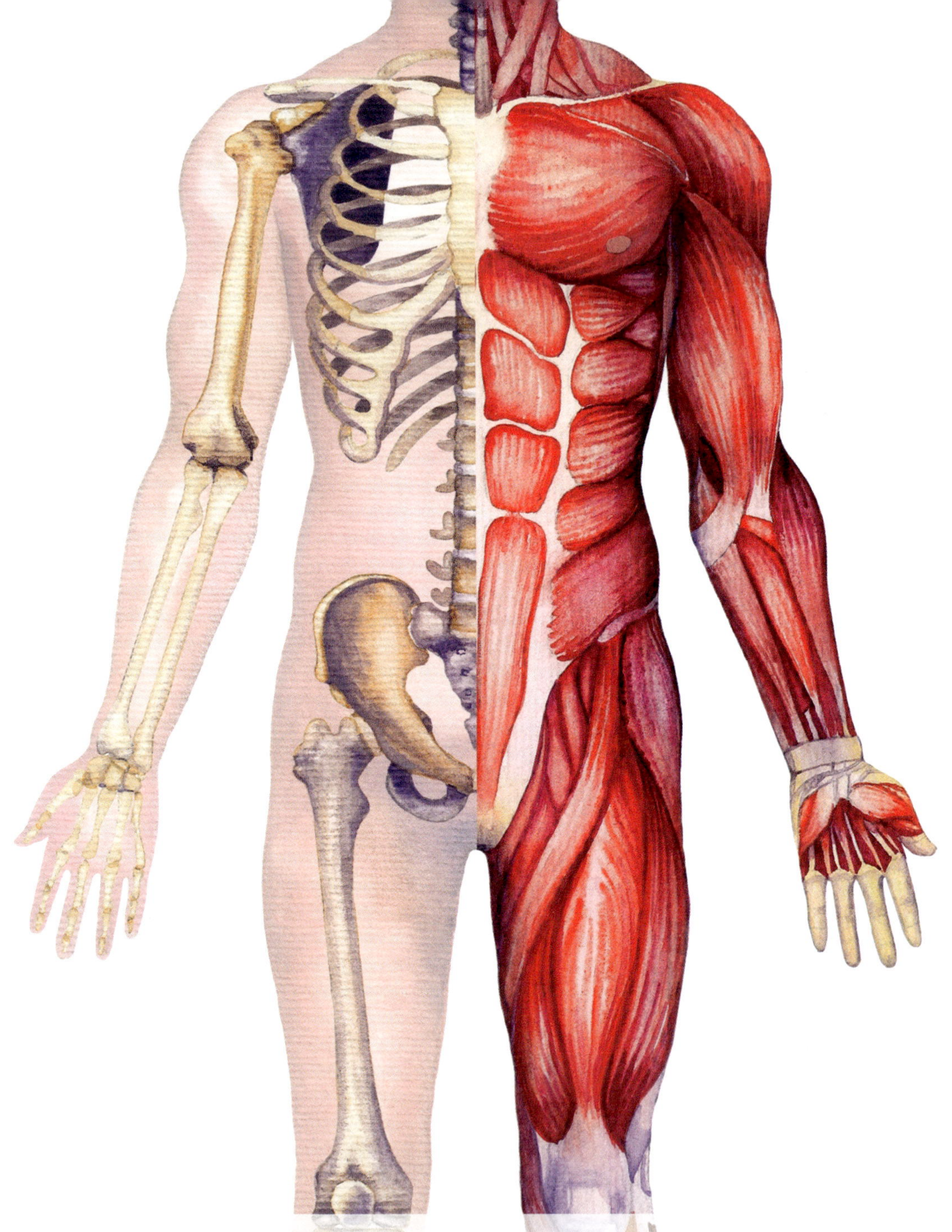

Your muscles and skeleton work together to help you move.

MOVEMENT AND SUPPORT

The nervous system sends its messages to the body, but where do they go? Motor neurons produce a chemical that "speaks" to muscle fibers. It tells the muscle to contract, or shorten. Then, the muscle moves.

When skeletal muscle contracts, it may move the nearby bone or bones. This is how you walk, talk, ride a bike, and complete any other voluntary action in your day. Let's look at how the muscular and skeletal systems work together to allow movement and give support to the body.

The Muscular System

Muscles are amazing body parts. Made up of thousands of small fibers, muscles are stretchy like elastic. They allow movement in the body when they contract and relax.

> **Fast Fact**
>
> Skeletal muscles and bones make up the musculoskeletal system.

Three different kinds of muscles make up the body. Smooth muscles carry out involuntary movements. They're found in your digestive system, helping food make its way through the digestive **tract**. Cardiac

The biggest muscles in your body are in your lower body. You can feel these muscles work when you climb stairs!

Fast Fact

The human body has more than 600 different muscles!

muscle, sometimes called myocardium, also handles involuntary movements. This is the muscle of the heart. Its contraction and relaxation moves blood in and out of the heart.

Skeletal muscles control voluntary movements like jumping, bending, and running. To keep your muscles healthy, stretch before exercising, and drink plenty of water.

The Skeletal System

Without bones, your body would have no structure! You wouldn't be able to stand or sit up straight. Bones protect your body's most important

TENDON TALK

Skeletal muscle connects to bone by cord-like tissue called tendons. When a muscle contracts, the attached tendon moves. The bone the tendon is attached to also moves. Tendons are very strong and flexible to transfer the movement from muscle to bone. They're found in places such as your elbow, wrist, shoulder, and knees. The largest tendon is the Achilles tendon, which connects your heel bone to your calf muscle. Tendons are tough, but they can be damaged. Tears and overuse can make it hard to do certain activities that use a certain tendon. For example, someone who ruptures their Achilles tendon would not be able to walk much until it healed.

Inflammation of the tendons causes tendinitis. This often happens in the knee, elbow, or Achilles tendon.

organs. The ribs protect the heart and lungs. Vertebrae, or the spinal column, protect your spinal cord. Without these protective bones, the most important organs in your body would be open to injury.

Two kinds of tissues are found inside bones. Cancellous bone looks like a sponge. This is where **bone marrow** is and where blood cells are made. Compact bone is the hard outer layer. It is very strong, and not easily broken. The largest and strongest bone in the body is the femur, which is found in the thigh. It can support up to 30 times a person's body weight! To keep your bones healthy and safe, get plenty of calcium in your diet, and wear a helmet and other protective gear when riding a bike or doing certain sports.

Musculoskeletal Issues

Issues with the musculoskeletal system are very common but very hard to deal with. That's because this system affects how a person moves. Musculoskeletal disorders include issues with muscles, bones, and joints. They often involve pain and mobility, or movement, problems.

The cranial bones protect your brain. Along with your facial bones, these make up your skull.

Fast Fact

Bones connect to other bones at joints. Strong, stretchy bands of connective tissue called ligaments surround the joint and connect the bones.

When you break a bone, it's called a fracture. Some fractures are fixed through surgery, while others just require immobilization.

The World Health Organization says that musculoskeletal conditions are a leading cause of disability worldwide. People with mobility limitations and pain may find it hard to work, take care of themselves and others, and do the things they once enjoyed. Neck and back pain are common musculoskeletal conditions. Another musculoskeletal condition, rheumatoid arthritis, causes inflammation in the body, especially the joints, that can make it hard to do certain actions.

Animals and humans need oxygen to live. Oxygen is released by plants through photosynthesis.

BREATH AND BLOOD

To check if someone is alive, you see if they're breathing and check for a pulse. That shows just how essential the respiratory and circulatory systems are! The respiratory system is responsible for taking in oxygen from the air. The circulatory system uses blood to carry oxygen to every cell of the body.

These two body systems make a great team. Together, they are responsible for sending essential resources—oxygen and blood—where they need to go in the body. This keeps cells alive, healthy, and ready to perform important functions.

The Respiratory System

A person takes about 20,000 breaths every day. Breathing in is called inhalation, breathing out is called exhalation, and the two functions together are called respiration. Though breaths last only a second or two, this important function supplies the oxygen that bodies need to survive. In fact, most people would die or have serious brain damage after about four minutes without oxygen.

Fast Fact

The lungs, heart, and bronchial tree are all found in the chest cavity, or thorax.

Taking deep breaths in and out can calm the nervous system when you feel upset.

Air comes in through the nose or mouth, known as the airway. The two paths of the airway meet at the throat. The breath continues down to the windpipe, or trachea. At the bottom of the trachea is the bronchial tree. It consists of air tubes called bronchi that connect to the lungs, smaller tubes called bronchioles within the lungs, and alveoli. Alveoli are very small sacs where oxygen and carbon dioxide are traded. You breathe in oxygen, then breathe out carbon dioxide—a harmful waste product. To keep the respiratory system healthy, never smoke, and wash your hands often to keep certain respiratory illnesses away.

Fast Fact

Some respiratory infections lead to bronchitis, or the inflammation of the bronchial tubes' lining. This makes it hard to breathe.

The common cold is one kind of respiratory infection, caused by a virus. It can give you a clogged nose, cough, and sore throat.

The Circulatory System

Oxygen is transferred, or sent, to the cells of the body by the circulatory system. It's carried in the blood. Every heartbeat carries oxygen-rich blood to the body's cells. When the old blood returns to the heart, the heart redirects it to the lungs. The respiratory system provides oxygen once again. This happens over and over, for the whole life of a human being.

The star of the circulatory system is the heart. It contracts, sending oxygen-rich blood out through the arteries. (The largest artery is the aorta.) It relaxes, allowing oxygen-poor blood back in through the veins. The heart has four chambers, or parts, that have

DISEASES AND DISORDERS

Conditions that impact the blood vessels or heart are common and can be serious. Heart disease is the leading cause of death in the United States among adults. Diseases of the circulatory system can cause problems with the pumping of the heart and issues with blood flow. Some people have arrhythmias, or irregular heartbeats, that prevent the heart from pumping blood as it should. Some people have **blood pressure** that is too high or too low. High blood pressure, also called hypertension, can lead to strokes, heart attacks, and heart failure. It's important that people with blood pressure problems go to the doctor for treatment.

Heart attacks happen when blood flow to the heart is blocked. Some are caught early enough and cause little damage, while others are deadly.

BLOOD FLOW THROUGH THE HEART

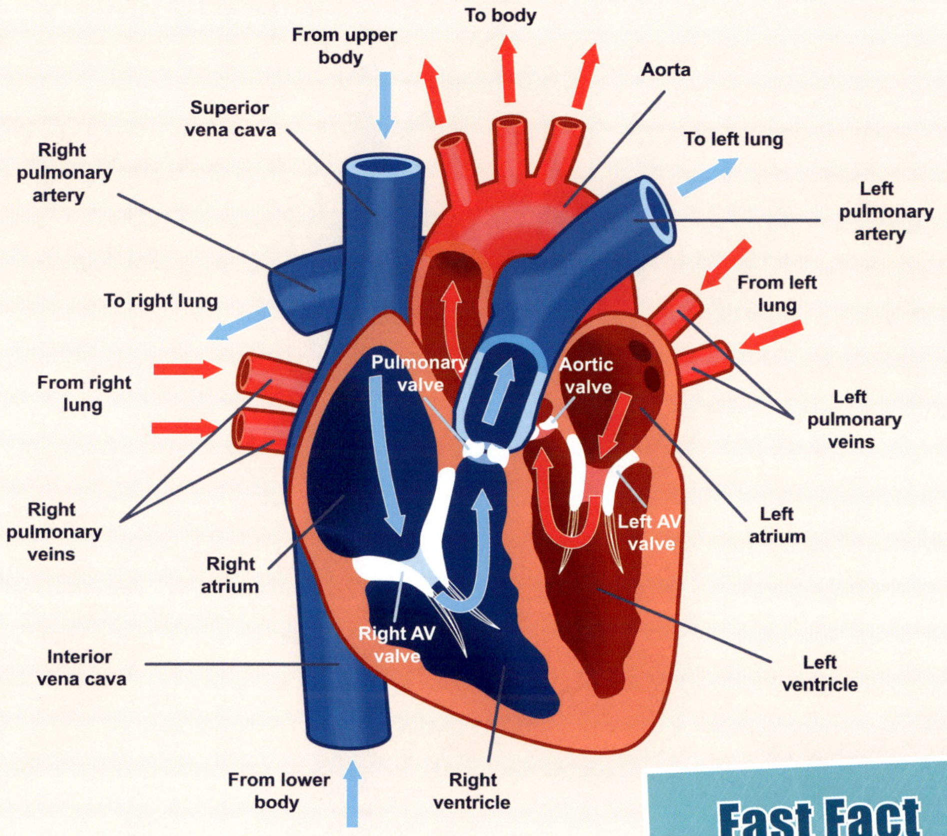

Follow the path of blood flow to and from the heart.

Fast Fact

The heart beats about 60 to 100 times every minute.

specialized jobs for pumping blood in and out. Blood vessels act as highways for the blood throughout the body. Blood brings hormones, nutrients, and oxygen to the body's cells. The branches of blood vessels decrease in size from the aorta to tiny vessels called capillaries in your fingertips. To keep your circulatory system healthy, you can eat nutrient-rich foods like whole grains and lean proteins, move your body throughout the day, and get plenty of rest.

The digestive system breaks down foods such as these so the body can get nutrients.

IN AND OUT

The body needs oxygen, but it also needs nutrients. We get our nutrients from the food we eat. The different nutrients, from proteins to vitamins, keep the body running.

The digestive system breaks down the food you take in. The moment you start to chew a juicy strawberry or a crunchy carrot stick, the digestive system gets to work to gather as many nutrients from the food as possible. This fuels the body and gives it energy to perform important functions through all the other body systems.

What Goes In

The mouth is the beginning of the digestive system. When you feel hungry, your body makes saliva, or spit. When you put food in your mouth, the saliva gets to work. It breaks down the food's chemicals. You swallow the food and it goes to the esophagus, a pipe that connects your throat to your stomach.

Fast Fact

Many "good" bacteria live in the gut and help break down certain foods. Your body has more bacteria than human cells!

Your teeth, tongue, and saliva work together to break down food in the mouth.

The muscles in the esophagus move the food along. When it enters the stomach, gastric juices break the food down into a liquid. The food stays in the stomach for an hour or so, and then it moves on to the small intestine. This organ breaks down the food further, absorbing vitamins, minerals, and other nutrients that are good for your body. It moves on to the wider, yet much shorter, large intestine. This organ absorbs the water and any final nutrients from the food, so the waste that's left behind is ready to come out.

The small intestine is much longer than the large intestine. On average, it is 22 feet (7 meters) long!

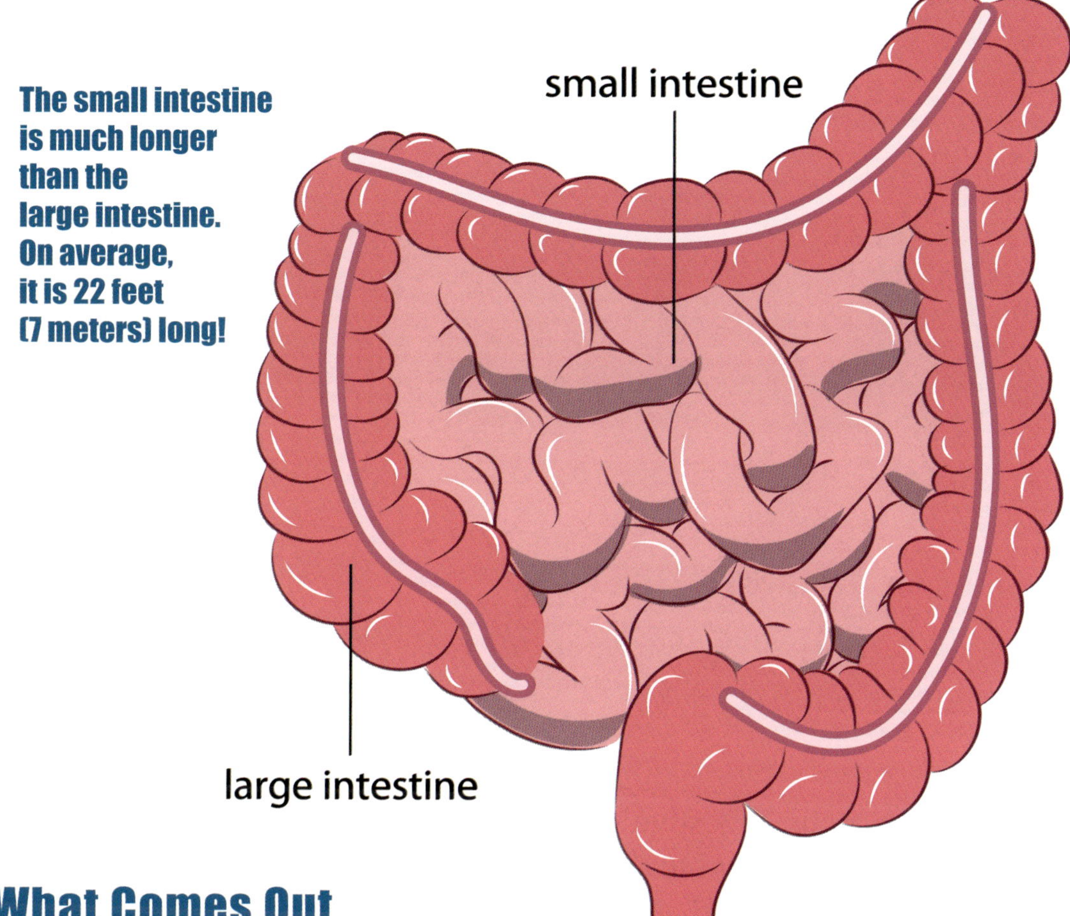

What Comes Out

By the time your food reaches the end of the large intestine, the useable parts have been taken away. The rest is a waste product called feces. Solid waste stays in the rectum at the end of the large intestine until you have a chance to go to the bathroom. Getting rid of the body's solid waste through feces and liquid waste through urine keeps the body healthy. It rids the body of things that are not necessary.

Fast Fact

The kidneys take waste products from the blood and make liquid waste called urine.

THE ENDOCRINE SYSTEM

The endocrine system has an important job in the human body. The system is made up of glands, which release specialized hormones into the bloodstream that help control growth, reproduction, organ function, and **metabolism**. Major glands in the human body include the adrenal glands, pancreas, pituitary glands, and thyroid. The pituitary gland is found in the brain. This pea-size gland is called the "master gland" because it directs other glands. It makes growth hormones, as well as hormones that direct the kidneys and thyroid. The kidneys balance water and waste in the body. The thyroid gland regulates metabolism and energy use in the body.

To keep your digestive system running smoothly, it's important to drink lots of water. You can also eat food that's high in fiber. Choosing foods with different nutrients will allow your body to grow and fight illness.

Fast Fact

The immune system's job is to fight infection from harmful "invaders" such as bacteria and viruses.

Systems Work Together

There are 11 organ systems in the body, including those that haven't been covered yet—such as the urinary, reproductive, and immune systems. Each body system has an important job to do within the body.

These systems also work together as a team. For example, nutrients from the digestive system aid the endocrine system and the immune system. Deep breaths from the respiratory system can calm

The Body as a Team

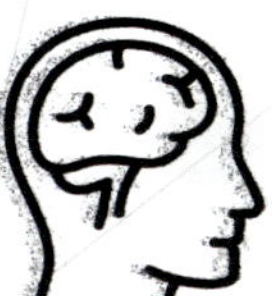

Nervous System: The Control Center

directs both voluntary and involuntary actions and processes in the body

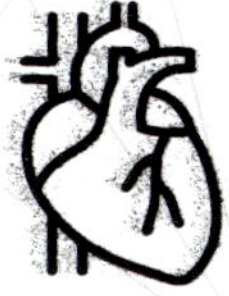

Circulatory System: The Blood Pumper

pumps oxygen-rich blood to the body and gets rid of oxygen-poor blood through the lungs

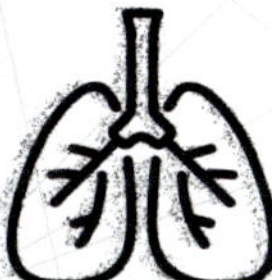

Respiratory System: The Oxygen Provider

takes in oxygen and delivers it to the heart

Digestive System: The Food Transformer

breaks down food into nutrients the body can use

Muscular System: The Mover

makes involuntary and voluntary movements

Skeletal System: The Supporting Cast

protects important organs and gives the body structure

The human body is like a team, and each system is like a player on that team. Just like every player is an important part of a team, every system is an important part of the human body. These are just some of the body's many systems!

the nervous system. The respiratory system takes in the oxygen that is used by the circulatory system to send oxygen-rich blood to the body's cells.

It's important to do your part too. You can be a team player by taking care of your body and its systems in the ways shared in this book. It's smart to treat your body with kindness and care.

THINK ABOUT IT!

1. How do you think mental health affects the body's systems?

2. Why is it important to rest the brain and body?

3. Why is it important to stretch before exercise?

4. What are some ways you take care of your body and its systems?

GLOSSARY

blood pressure: The force of blood pushing on the walls of blood vessels that varies with physical condition and age.

bone marrow: A soft substance inside bones where blood cells are made.

complex: Having many parts.

coordination: When parts work well together for effective results.

hormone: A product of living cells that circulates in body fluids (such as blood) and produces an effect on the activity of other cells.

immobilization: Taking away the ability to move or function.

inflammation: A local response to cell injury that causes redness, heat, and pain.

metabolism: The chemical changes in living cells by which energy is provided for vital processes and activities.

specialized: Produced and made for a specific purpose.

tract: A system of body parts or organs that act together to perform a function.

voluntary: Something that is done from one's own choice or will.

FIND OUT MORE

Books

Barnes, Kate. *The Body*. Buffalo, NY: Cavendish Square Publishing, 2025.

McClure, Leigh. *The Nervous System*. Buffalo, NY: Scientific American Publishing, 2025.

Walker, Richard. *Human Body*. New York, NY: DK Publishing, 2023.

Websites

15 Facts About the Human Body!
www.natgeokids.com/uk/discover/science/general-science/15-facts-about-the-human-body/
Discover incredible facts about the human body that will blow your mind!

The Human Digestive System
www.natgeokids.com/uk/discover/science/general-science/your-digestive-system/
Follow the path your food takes through your digestive system.

Your Brain & Nervous System
kidshealth.org/en/kids/brain.html
Learn more about the nervous system and its star—the brain.

Publisher's note to educators and parents: Our editors have carefully reviewed these websites to ensure that they are suitable for students. Many websites change frequently, however, and we cannot guarantee that a site's future contents will continue to meet our high standards of quality and educational value. Be advised that students should be closely supervised whenever they access the internet.

INDEX